# Cities and Floods: A Pragmatic Insight into the Determinants of Households' Coping Strategies to Floods in Informal Accra, Ghana

Kwaku Owusu Twum
Mohammed Abubakari

ELIVA PRESS

ELIVA PRESS

## Kwaku Owusu Twum

## Mohammed Abubakari

This book examines the major effects of perennial flooding on households living in Alajo, an informal suburb of Accra, the national capital of Ghana. In furtherance, it considers the factors that determine the coping strategies of households and then delve into the diverse coping strategies that have been adopted by the informal urbanites which have influenced their continual stay despite facing the risks of flooding every year.

Floods are common events that confront many cities in the developing world. Ghana, a developing country, is persistently challenged with flood events, especially in its major cities. In informal Accra for instance, despite the severity of flood effects and its associated threats, poor informal residents continue to stay. As a result, these poor urban dwellers have developed local coping strategies made up of mitigation and reactive measures to manage and adapt to flood hazards through their preceding experiences. In this paper, we have embraced the Convergent Parallel Mixed Method of Case Study Design to echo and explore: (i) the major effects of preceding floods on informal households; ii) the local informal coping strategies adopted by households to mitigate and respond to flooding and its effects in the future; and iii) the determinants of the coping strategies of households which underpin their continual stay in spite of flood risks in Alajo, an urbanized suburb in Accra metropolis noted as one of the slum communities that easily floods in Ghana. Our analysis has used a mix of qualitative and quantitative data collected from both secondary and primary sources as well as a conceptualized model known as Disaster Resilience of Place (DROP). The key finding (Alajo has low degree of adaptive resilience to major floods which might occur in the future due to the lack of social learning in the coping strategies developed through several years of lessons learnt from perennial floods) and proposals (local coordination in implementing the coping strategies to flooding, state support of the local strategies and adoption of rainwater harvesting) also make contributions to managing urban floods in informal settlements in the developing world.

Published: Eliva Press SRL

Address: MD-2060, bd.Cuza-Voda, 1/4, of. 21 Chişinău, Republica Moldova

Email: info@elivapress.com

Website: www.elivapress.com

ISBN: 978-1-63648-015-2

# Contents

## ABSTRACT

Floods are common events that confront many cities in the developing world. Ghana, a developing country, is persistently challenged with flood events, especially in its major cities. In informal Accra for instance, despite the severity of flood effects and its associated threats, poor informal residents continue to stay. As a result, these poor urban dwellers have developed local coping strategies made up of mitigation and reactive measures to manage and adapt to flood hazards through their preceding experiences. In this paper, we have embraced the Convergent Parallel Mixed Method of Case Study Design to echo and explore: (i) the major effects of preceding floods on informal households; ii) the local informal coping strategies adopted by households to mitigate and respond to flooding and its effects in the future; and iii) the determinants of the coping strategies of households which underpin their continual stay in spite of flood risks in Alajo, an urbanized suburb in Accra metropolis noted as one of the slum communities that easily floods in Ghana. Our analysis has used a mix of qualitative and quantitative data collected from both secondary and primary sources as well as a conceptualized model known as Disaster Resilience of Place (DROP). The key finding (Alajo has low degree of adaptive resilience to major floods which might occur in the future due to the lack of social learning in the coping strategies developed through several years of lessons learnt from perennial floods) and proposals (local coordination in implementing the coping strategies to flooding, state support of the local strategies and adoption of rainwater harvesting) also make contributions to managing urban floods in informal settlements in the developing world.

3

# 1. INTRODUCTION

Over the past few decades, urban communities have become the central hub for human existence. People prefer to live in the urban areas instead of the rural communities. An estimation made by the UN-Habitat (2010) tends to suggest that about 70% of world's population will live in the urban centers by 2050.Even before 2050, the world should expect more than 60% of its population living in urban areas by 2030, with Africa recording a rapid rate of urbanization (UNCHS, 2007; Adegun, 2011). The rising spate of urban share of population has led to the development and continual expansion of informal settlements and slums (UN Habitat, 2010) which are susceptible to disasters such as flooding (De Risi, 2003).

The frequency and severity of flooding in African cities have been heightened (Douglas et al., 2008) ranking second to Asia (Tschakert et al., 2010) which has the highest rate of urbanization in the world. In Ghana, for every ten (10) people, more than four (4) live in the urban centers (Songsore, 2009), with over half (58%) of the urban population living in slums (as indicated in Growth and Poverty Reduction Strategy, 2006-2009). The slum dwellers are persistently confronted with floods, especially in Accra, where within the period of 1995-2007 alone, more than ten (10) floods were recorded characterized by deaths, households' displacements and huge loss of infrastructure, properties and capital (Aboagye, 2012). Rapid urbanization, going hand in hand with increasing concentration of human activities, densities and congestion, pollution and impermeable surfaces without commensurate measures have increased exposure and vulnerability to flooding triggered by heavy rainfall, and these have been common contributions to flood events in Africa, including Ghana (Douglas et al., 2008). Ghana's urbanization is widely disorganized, and as such the millions of desperate and optimistic rural migrants who enter the cities have no appropriate place to stay and/work. In order to fit into the cities, these groups resort to informal locations, usually flood prone areas that constantly put their lives and properties under

4

serious threat (Satterthwaite 2007; Douglas et al., 2008; Jha et al., 2012). The frequency of floods in informal areas in Ghana is enormous, with severe negative implications on the poor and marginalized people vis-à-vis individuals living in formal locations (Owusu and Afutu-Kotey, 2010). The informal areas are inundated centers which are industrially 'rejected', and are characterized by unsecured slopes, high levels of pollution and hazards (Satterthwaite et al., 2011). These are significant features which have made the economic values of informal areas considerably very low, thus serving as attraction points and preferred destinations for the urban poor amidst the dangers posed for staying there. These poor urbanites are more concerned about the economic gains rather than the threats and risks they are exposed to. As such, they live as 'survivors' under the mercy of flood disasters, where their lives and properties are continuously threatened. Although this may seem irrational, the informal population continue to rise and their influence consolidates progressively. In Kumasi, the second major city in Ghana, Adjei-Mensah et al. (2013) revealed strikingly that more that 90% of informal dwellers had neither thought of moving away from their informal settlements (91.6%) nor made plans to live in formal settlements (93.7%). This revelation is quite surprising, but these people live in such areas as they have no option due to their limited resources, and as such they are forced to live and perceive such areas as 'right place of abode' for their existence.

In managing floods in Ghana, city authorities have commonly embraced the issuance of eviction notices and various evacuation threats to informal residents (Owusu and Afutu-Kotey, 2010); an approach that has proven unsuccessful but is constantly adopted as if there are no alternative ways of controlling urban informality. In the phase of floods, local authorities tend to give more priority to commercial and administrative areas whereas the poor informal communities, which are the most affected centers, are less prioritized (Douglas et al., 2008). This has therefore induced informal dwellers living in flood prone areas to

5

develop local mitigation measures and reactive responses to manage and adapt to flood disasters (Amoako, 2014). They are able to sustain flood event to some degree despite the severity of its impacts and in effect, has caused them to be adamant to the threat of city authorities. These local mitigation measures and responses have been shaped from the incremental learning and the determinant factors including topography, locational disadvantage and absence of state support. This research has used Alajo, an informal settlement in the Accra metropolis of Ghana, as a case study. First, it has examined the preceding effects of flooding on the informal households in the community. This is the followed by the exploration of how these households prepare and respond to flooding, and what has influenced these measures used to mitigate and respond to the flooding. The findings have been discussed conceptually using the Disaster Resilience of Place (DROP); a model developed by Cutter et al. (2008) for place-specific disaster studies.

# 2. UNDERSTANDING URBAN INFORMALITY AND FLOOD, FLOOD VULNERABILITY, FLOOD HAZARDS AND FLOOD COPING STRATEGIES FOR LOCAL FLOOD MANAGEMENT

This chapter gives an overview of urban informality and floods, flood vulnerability, flood hazards and flood coping strategies as a way to build local adaptation and resilience to flood hazards.

## 2.1. Urban informality and floods: an overview

Urban informality is a popularly discussed issue in urban literature. It is thus, predominantly known and persistently explored by urban planners and development partners. Scholarships on urban informality continue to evolve as an attempt to unravel its complexity and to find responsive measures to appropriate it into urban development interventions. De Soto (2000) has extensively studied urban informality through an economic lens. He subsequently defined informal economy as "people's spontaneous and creative response to the state's incapacity to satisfy the basic needs of the impoverished masses" (as in The Other Path, 1989:14). Subsequently, through continuous studies of informality, he presented it as "heroic entrepreneurship", where the poor urbanites earn their source of income for existence. Whilst Hall and Pfeiffer (2000) generally perceive urban informality as an urban development problem, Roy and Al-Sayyad (2004) have argued that informality in itself should not out-rightly be tagged as a problem, rather a process that governs how urban communities are transformed through a series of transactions that connect formal and informal spaces and economies to one another through a system of norms and logic. In the phase of disparity in understanding and presenting urban informality, characterized by the lack of consistency in both theoretical and empirical research (Guha-Khasnobis et al., 2006), the choice of how to depict urban informality should be influenced by the context of its discussions (Heintz, 2012). Following

7

discussions, urban informality can generally be divided into two components: informal settlements and informal economic sector (Porter et al, 2011). Thus, the lack of consensus is deepened by the angle in which scholars tend to view the term. In this paper, we view urban informality in a similar lens as Duminy (2011), who presented it as series of behaviors and practices evolving within cities, which are relatively unregulated or uncontrolled by the state/formal institutions. Informality is widely varied, as in terms of housing and jobs, and it could offer an avenue not only for the informal dwellers, but also, the formal residents in the urban centers (Werna, 2001). Thus, informality is not merely compose of socio-spatially marginalized people, but also, well-to-do individuals who work within the formal sector and/or individuals who live in formal communities but work within the informal sector.

The UN-Habitat (2003) aligns informality to informal settlements by clearly indicating that informality is closely associated with issues of illegal occupations on lands, and unauthorized houses and structures; a phenomenon very common in the developing world. This is explained by the fact that emerged/emerging physical developments do not meet planning approval and structures are not in conformity with building and zoning regulations (Fekade, 2000). Due to the nature of informality in the developing world, Elgin and Oztunali (2012) have contended that urban informality is rife and has serious social, political and economic development challenges. The rise in disasters, notably, flooding and the spread of overcrowding, congestion, crime and theft has direct connection to urban informality (Twumasi-Ankrah, 1995). Paying attention to informality and floods, Dodman, Bicknell and Satterthwaite (2012) agree that informality contributes to flood events. This comes about when a high proportion of lower income groups settle on hazardous sites (sites at risk of floods or landslides), which they do so for the lack of safer lands which meet their economic standards of affordability. The dwellers of such areas are vulnerable to flood and other

8

hazards due to their huge concentration and densification of businesses within the same locations (Douglas et al., 2008). This is further worsened by the poor road access to informal communities which serve as a great obstacle for easy exit during emergencies. The increase in concentration leads to the expansion of informal areas into nearby formal communities and occupation of marginal lands on the urban periphery (Satterthwaite et al., 2011). This is incrementally manifested through the alteration of natural landscapes, land uses and land cover, which add to flood hazard problems. This is because, apart from the alteration of natural landscapes, land uses and land cover, informal practices such as insufficient removal of refuse and drainage infrastructure for available population are transcended to such nearby formal areas posing such areas to flood risks. Sakjege, Lupala and Sheuya (2012) expatiate on the contribution of informality to flood events by indicating that as densification increases, the run-off water from the roofs of buildings alters the urban land cover and land surface, and this is further exacerbated by poor solid waste management practices of informal residents. Thus, floods are caused by not only natural circumstances but, human's activities as well (Douglas et al., 2008). As climate change influences flood events (Amoako, 2014), so as local urban change increases the likelihood of floods due to the adjustments to the urban land surface and water passage-ways as a result of human activities such as structural developments, flooring or paving (impervious surfaces), soil compaction, the elimination of vegetation and the digression of natural flows. Subsequently, these informal areas play a major role in the cause and severity of flood effects.

## 2.2. Flood vulnerability, Flood hazard and coping strategies for building community-level resilience to floods

### 2.2.1. Flood vulnerability and Flood hazard

In order to understand flood vulnerability, the term 'vulnerability' needs to be explored and appreciated. Though the true origin of 'vulnerability' can be traced from the field of geography and natural hazards, many other related disciplines now use the term in a contextualized manner (O'Brien et al., 2004a; Gow, 2005). Others relate the term to concepts such as marginality, susceptibility, resilience, fragility, and risk (Liverman, 1990), as such making it complicated to define. Hence, Kasperson et al.(2005) concluded that a single accepted definition for vulnerability is non-existence. Brooks (2003) has also reiterated that one can meaningfully define vulnerability based on specific hazard(s) of a particular system so as to differentiate between current and future vulnerability. On the basis on the ambiguity in meanings to vulnerability, we contextualize our definition of vulnerability in tandem to flood events in informal urban communities. We adapt Turner II et al. (2003) general definition of vulnerability as "the degree to which a system is likely to experience harm due to exposure to a hazard" and contextualize it vis-à-vis urban informality and floods. Flood vulnerability is thus, defined as: the degree to which informal urban communities are likely to experience harm due to their exposure to flood hazards. We further define flood hazards based on the adapted definition of hazard by the United Nations (2004) as "a potentially damaging physical event, phenomenon or human activity that may cause the loss of life or degradation". On this premise, flood hazard is defined in this research as: potentially damaging flood event induced by nature or human activities that may cause the loss of life or injury, property damage, social and economic disruption or environmental degradation in informal urban settlements.

## 2.2.2. Flood coping strategies: An approach to building local adaptation and strengthening resilience to flood hazards

Before we look into the specific coping strategies of informal communities to flood hazards, it is prudent to look into the terms 'coping', 'adaptation' and 'resilience' and establish clearly what they stand for in this research.

Coping is made up of immediate and short-term measures to an event, culminating 'here and now' capacity of a system/community to mitigate and respond to the event (Birkmann, 2011). Adaptation is somewhat long term process which entails systematic approach of learning (either planned or spontaneous in nature), experimentation and change before or after a disaster (Yohe and Tol, 2002; see also Pelling, 2010). Resilience is the capacity of a system/community to prevent, mitigate and/or cope with risk, and recover from shocks (FAO, 2012). Resilience is a state informed by sound coping measures and adaptation, and take considerable period of time for community to reach. Carpenter et al. (2001) have revealed that resilience can be specified in the context of vulnerability as "resilience of what to what". A system could thus be said to be resilient when it is less vulnerable to shocks across time, and can recover from them. Communities can build resilience when they adapt to risk through incremental and social learning from adaptation measures.

In the context of urban informality and flood, we can infer that communities can systematically learn from their flood coping strategies (short term) which can inform their adaptation measures to flood hazards. Through this, informal communities can subsequently strengthen their adaptation to floods, helping them to reach a state of resilience to flood hazards over time (see Figure 1). Placing emphasis on coping strategies of informal communities to floods, Douglas et al. (2008) have indicated that the strategies are mostly individually initiated and disorganized, hence, making their contributions to building local adaptation, and subsequently helping communities to reach a state of resilience to flood questionable. These coping strategies could entail the construction of barriers to impede the inflow of water into houses during flood hazards, building temporal shelters (Adelekan, 2010); moving important items to safe places in the event of flood, creating high places in homes using furniture, stones and blocks where items can be kept temporarily during floods; and treating drinking water through boiling to reduce the risk of water-related diseases (Douglas et al., 2008). In a related study conducted by Sakjege, Lupala and Sheuya (2011), flood coping strategies at the household level were the use of sandbags and tree logs, water boiling and chemical treatment, raised doorsteps and pit latrines, construction of proactive walls and elevation of house foundations, temporal movement to safe places and provision of pipe outlets to drain off water during heavy rainfalls. The study further unraveled that some communal-based interventions such as control of housing development which tend to contribute to floods, protest, attempts to request companies which have contributed to flood to compensate community members and initiation of communal solid waste management practices have been implemented but were unsuccessful. In the light of the difficulties faced by informal residents in organizing themselves in implementing flood coping strategies for the entire community, Amoako (2012) has revealed that individual and household-based interventions have produced very little success in local flood adaptation and helping them to incrementally build resilience to flood hazards. Informal community members are somehow able to cope with flood events in the short term, but still have high risk to flood hazards despite flood coping strategies. This is due to their exposure to flood, as they still remain in areas where flooding occurs. Perhaps, the uncoordinated nature of local flood coping strategies among the informal settlers is a reason for their less recognition and support by state formal institutions.

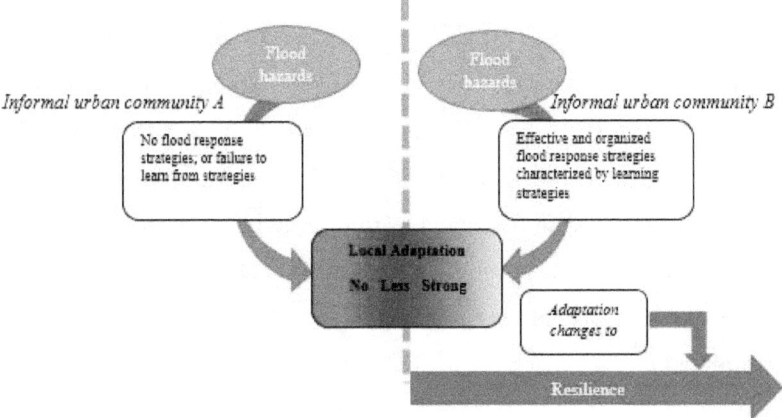

*Figure 1: Local flood adaptation and resilience*
**(Source: Authors' Construct, 2018)**

Figure 1 shows two informal urban communities A and B, with B having built resilience to flood hazards through strong adaptation ensured by the effective and organized flood coping/response measures characterized by learning strategies by the local dwellers. On the hand, community A has no/less adaptation due to either the lack of local flood coping/response measures to adapt to flood or community members have failed to incrementally learn from the measures put in place to inform adaptation helping them to build resilience over time.

## 3. Disaster resilience of place (DROP): ADOPTED MODEL FOR UNDERSTANDING INFORMAL HOUSEHOLDS' RESPONSES TO FLOODS

The research adopts the Disaster Resilience of Place (DROP) model developed by Cutter et al.(2008) as a place-based model for understanding how community develops resilience to natural disasters such as floods through adaptation. The DROP model explains, in the light of theory, how local communities build resilience to disasters through incremental learning, and practical mechanisms to which real life disaster problems can be addressed in real places. The model is underpinned by three critical assumptions which have justified why the researchers adopted it as a conceptual model for this research. They are:

i) The model can be used to study a wide range of disaster problems including floods, a perennial disaster in which this paper tends to explore in informal Accra;

ii) The model's applicability is at the local level instead of national and/or international, hence making it appropriate to be adopted; and

iii) The model focuses on the connection between social systems, natural systems and physical/built environment and how they contribute to; and be used to manage place specific disasters such as floods.

Place-specific factors between social systems, natural systems and the physical/built environment influence the degree of inherent vulnerability and inherent resilience (see nested triangle in Figure 2) of a particular community to disaster. These place-specific factors vary and could entail poverty, rapid population growth, disorganized buildings due to poor governance, socio-economic marginalization (Braun and Aßheue, 2011) and features of the landscape that increase exposure to hazards. The inherent process both in the context of vulnerability and resilience happens at the local scale and is influenced by the multi-scale factors interconnected by social systems, natural systems and the built environment (as in the bigger triangle). Cutter et al. (2008) treated the factors as 'antecedent conditions' which interact with the disaster event characteristics. Disasters then occur when people exposed to the hazards are vulnerable to their effects. The disaster event characteristics include the duration of the disaster, its number of occurrences (frequency), intensity, magnitude and rate of its onset. These immediate effects are either reduced through effective coping strategies (indicated with a minus (-) sign) or intensified through the absence of any strategy (indicated with a plus (+) sign). Thus, using informal communities as an example, households' use of local coping strategies to disaster-adaptation such as flood could help reduce the immediate effects of the flood. The strategies in the context of this study have been espoused in the findings section of the paper.

The disaster impacts are combination of the antecedent conditions, disaster event characteristics and the coping strategies embraced by local communities. The use of predetermined coping strategies by local communities in order to withstand the impacts of the occurrence of a known event/disaster underpinned the absorptive capacity of such communities. Absorptive capacity is the ability of a system (a community and/households) to survive shocks and stress—i.e. the system keeps functioning in the phase of disaster (Boubacar et al., 2017). Communities with effective implementation of appropriate coping measures will undeniably be able to reduce the impacts of an event/disaster should it occur and as well their absorptive capacity to the event/disaster will not be exceeded. At this point, the degree of recovery of the affected community will be high. On the other hand, the absorptive capacity of local communities to an event/disaster could either be exceeded when coping strategies are not sufficient to withstand the impacts of the event/disaster should it occur or the occurrence of the event/disaster will have severe impacts which exceed the coping strategies of local communities to the event/disaster. In the context of this study for instance, if households' coping strategies to flood events which occur perennially are not adequate or the occurrence of the flood events will be severe that its impacts will absorb the local coping strategies of households, then the absorptive capacity of the community will be exceeded. At this point, the community could perhaps, exercise adaptive resilience through immediate actions during the disaster and social learning to help in the recovery process after the disaster. Adger et al., (2005) have unraveled that social learning involves differences in adaptations as well as embracing and building strong social integration at the local level for group-based actions to recover from the occurrence of the disaster. The social learning becomes a process and feeds into future coping strategies for pre-event preparedness advancement which subsequently strengthens the absorptive capacity of communities for future events through inherent resilience indicated by the feedback loops in Figure 2.

It is worthy to note that once coping strategies are ineffective and local absorptive capacity is exceeded, households can learn some lessons (lessons learnt) in terms of what they did right and wrong, and this becomes future recommendations which could either be implemented or never implemented by individual households. This makes 'lessons learnt' quite apart from social learning which considers collective actions to deal with a disaster, which also feed into future strategies of households. Through social learning (collective initiative based on shared experiences and resources), with/without implementation of lessons learnt (informed by individuals' experiences and means to deal with a disaster), households are able to build adaptive resilience. Once adaptive resilience is built, households can successfully cope with risk, and recover from shocks associated with disasters. The degree of recovery for such households becomes very high.

Alternatively, a low degree of recovery occurs when after absorptive capacity has been exceeded, 'NO' (as indicated in Figure 2) adaptive resilience is built by the community. This occurs when coping strategies are either in non-existence or are very poor. Building resilience is very essential as new knowledge is gained through the process which provides feedbacks to modify the new coping strategies. The existence of social learning and implementation of lessons learnt in the entire process provide a potential platform to enhance preparedness (+) and mitigation (+) otherwise preparedness and mitigation will be affected negatively (-). The DROP model provides useful insight to community based disaster studies, but one limitation we identified is the failure of proponents to admit that 'external supports' can help in recovery process of communities whose absorptive capacity is exceeded as a result of ineffective and unsound coping strategies. The model thus, attempts to outline how vulnerable communities can become resilience over time through local measures without due recognition of 'external supports', especially from responsible state officials. Cutter et al.(2008) who designed that model have also called for the need for further studies which aim at advancing the model to pay credence to the development of guidelines that recognize structural, economic, social and environmental policy changes.

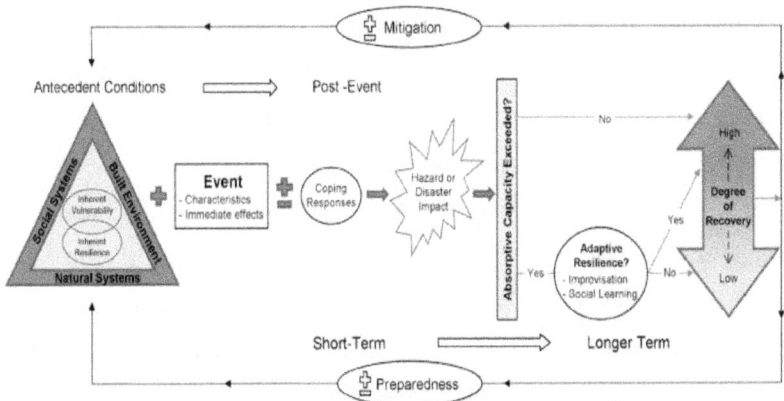

*Figure 2: The Disaster Resilience of Place (DROP)* **Model**
***(Source: Adopted from Cutter et al., 2008)***

17

# 4. STUDY CONTEXT AND METHODS

This chapter focuses on the profile of the study area as well as methods and materials used for the conduct of the research.

## 4.1 Study context

Alajo is one of the rapidly urbanizing centers in informal Accra. It is located within the Ayawaso Central Sub-Metropolitan area of the Accra Metropolis of Ghana (see the land use map of Alajo in Figure 3) and is only six (6) kilometers from the Central Business District (CBD) of Accra. The physical area of the community is relatively small, approximately one (1) kilometer sq. The main access road to the community is the Alajo road; a second class road, which has been encroached by small business activities, and jammed by vehicular traffic and pedestrians' movements to the CBD. There are small open spaces for recreational activities, and the entire community relies on a school-yard popularly called the '*polo park*' as a place for social gathering and engagements. The Accra Metropolitan Assembly (AMA) classifies Alajo as a third class income zone (AMA Medium Term Development Plan, 2014) which is the least for such classification, obviously suggesting that poverty levels are really high and there are no indications that the residents can afford rent in formal Accra. The land in Alajo is less than 50 meters above sea level which exposes it to inundations. The community is also located at the confluence of two main rivers, the Odaw and one of its tributaries, river Onyasia, which easily spread out when there are heavy downpours (Refer to Figure 4 and 5).The Odaw river has a watershed of about two hundred (200) kilometer sq. and Alajo is located toward the downstream extent, which is about five (5) miles upstream from Korle lagoon; the point where the Odaw river empties into the Gulf of Guinea. Residents who are closer to the rivers (100-150 feet) are within the flood risk zone identified by the UN-Habitat (2011) which shows the exposure of households to flood (Figure 5 and 6).

Impliedly, these people are very liable to floods, but have developed coping strategies to manage and adapt to flood events.

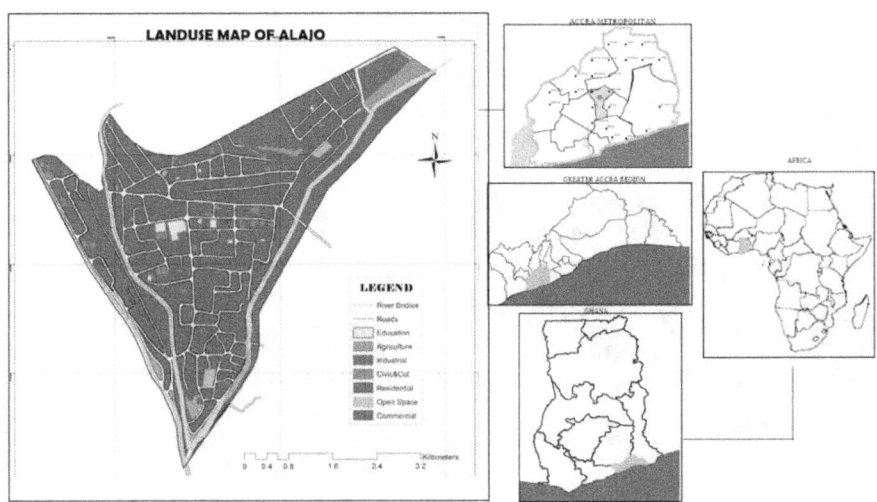

*Figure 3: Map of Alajo showing the various land uses*
**(Source: Authors' construct, 2018)**

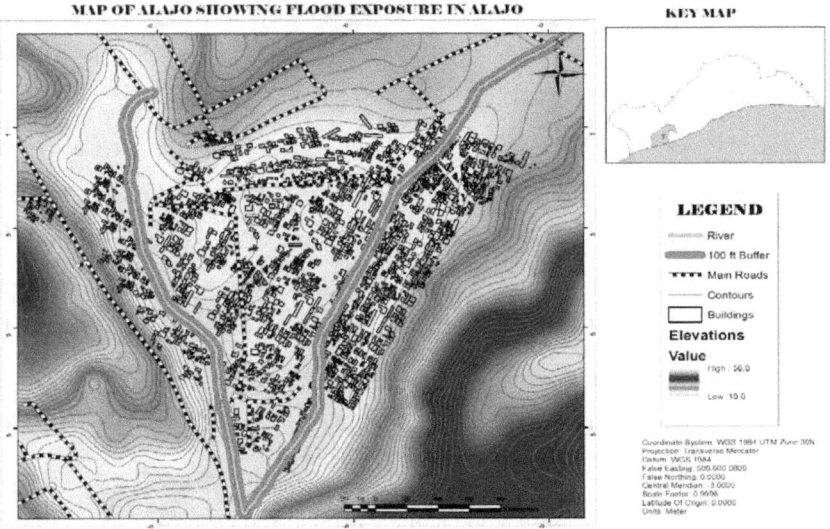

*Figure 4: Map of Alajo showing areas of flood exposure*
**(Source: Authors' construct, 2018)**

19

*Figure 5: Degree of flood exposure of Alajo using Triangular Irregular*

*Network (TIN)*

**(Source: Authors' construct, 2018)**

## 4.2 Materials and method

The research adopted the Convergent Parallel Mixed Method of Case Study Design. This design as indicated by Yin (2003) is flexible and relevant when a detailed and concentrated analysis of a single case is selected for a research. The design also allows data gathering from several sources including interviews, questionnaire administration and observation. The Convergent Parallel Mixed Method allows a more thorough understanding of problems through comparisons of diverse perspectives drawn from quantitative and qualitative data (QUAN + QUAL) (Morse, 1991). Again, this method of Case Study Design was used as it allows in understanding results and changes needed for a marginalized group by merging their inputs and allowing their perspectives to stand out in a research

20

(Hollohan and Barry, 2014). The Convergent Parallel Mixed Method can be embraced in a research by designing both quantitative and qualitative strands of research and analyzing them in a concurrent manner (Wittink et al., 2006; Creswell, 2012; Creswell, 2014). This has be done, but more emphasis has been granted to qualitative data in relation to quantitative data as an attempt to throw more light to the personal revelation of study participants (as suggested by Hollohan and Barry, 2014).

The community-level focus, choice of Alajo, is seen as very appropriate case for study due to its flood history and marginability, thus, we have explored the linkages between the various facets of its informality (housing conditions, livelihoods, legal status and rights), its flood vulnerability and coping measures. Data were extracted from both secondary and primary sources. A blend of both sources gave different accounts of already existing studies for comprehensive analysis. Secondary data sources were reports of preceding research studies, articles and journal papers and government publications. On the other hand, primary data were obtained through field study using interview guides, questionnaires and direct observation. Collection of data through questionnaires took the form of typed and printed predetermined set of questions which were administered to respondents to determine their perspective on issues related to the subject matter. This was done by the lead author of the research in 2015/2016 as part of his bachelor's thesis, who discussed with respondents wide range of issues convering informality and floods (including their choice of the community, the economic activities they engage in, number of years and experiences with flood, how they have been able to cope with flood, and suggestions they have for flood management).

Alajo lies between two major water bodies (the Odaw and Onyasia rivers) which are recognized as the major cause of perennial flood in the community (see

21

Atuguba and Amuzu, 2006; and Attipoe, 2015), hence, households who are within and outside a hundred feet (100 ft) from the rivers were purposively selected. This selection is underpinned by the fact that properties including building are not expected to be developed within a hundred feet (100 ft) away from the rivers (Revised Planning Standard, 2010). Again, for the purpose of ascertaining the required data, households living in houses as well as shacks and kiosks that are along the rivers were considered. This to to establish the fact that Alajo has some areas which are not flood-prone, and not everyone in Alajo is also poor. The aforementioned criteria served as the basis for inclusion and exclusion of households. Households who qualified based on the criteria were considered, but data were collected from a conveniently selected sample of sixty (60) households. The settlement patterns in the selected parts of Alajo were scattered, thus, the use of probability sampling (systematic and/or simple random sampling) will have taken longer time and efforts as well as greater financial commitment for the research which was based on personal funding. Resultantly, convenient sampling (opportunity sampling) became a better option in economic sense to enrol households who were ready to participate in the study. As argued by Dörnyei (2007), convenient samping can be used where conditions surrounding a research is unfavorable for probability sampling to be used. In this case, convenient sampling provides the avenue for researchers to engage field participants who are ready to provide their rich inputs deemed commensurate for drawing valid inferences and making solid analysis. In every house/structure visited, household heads were prefered, but in instances where household heads were unavailable, data were solicited from household members deemed legitimate to provide data to inform the research.

Apart from households, primary data were solicited from relevant institutions and stakeholders purposively selected including the National Disaster Management Organization (NADMO), the Accra Metropolitan Assembly (AMA) and the

Town and Country Planning Department (TCPD). Two (2) officers from each institution were selected and interviewed by the lead author. Thus, a total of six (6) state officials were engaged. This makes the total number of field respondents sixty six (66)---made up of sixty (60) household heads/reps and six (6) state officials.The data from primary respondents were supplemented with direct observational technique embraced by the researchers. Pictures were taken, the drainage patterns were studied to know the direction of flows and their effects on households and coordinates were also noted for derivation of contours in the study area. Even though the observations did not follow any organized procedure, they presented various key and relevant issues that were connected to achieve the objectives of the research. Data analysis was done both qualitatively and quantitatively. SPSS was used to assist in the analysis of the quantitative data whilst qualitative data were descriptively analyzed.

# 5. RESULTS AND DISCUSSIONS

This chapter captures the findings gathered for the study which have been presented within the realm of available literature on informality and flooding.

## 5.1 Major effects of the perennial flood events on the urban residents

The burgeoning studies on urban informality reveal the interconnection between informality and disasters and its associated effects on informal residents (see for instance; studies such as Twumasi-Ankrah, 1995; Satterthwaite, 2007; Elgin and Oztunali, 2012). Haphazard physical development characterized by concentration and densification of businesses is a common hallmark of informality (Fekade, 2000; Douglas et al., 2008). Alajo depicts such features, making the occurrence of floods very common. As a result, the occurrence of floods have produced some negative effects on huseholds (Table 1). The most significant effect has been the loss of household items. This was confirmed by majority of the selected households (68.3%). The disruption and loss of services and businesses by the informal dwellers was confirmed as another effect of the perennial flood. Not only that, households confirmed to outbreak of diseases in the community as one outcome of the perennial flood. These aforementioned factors were separately confirmed by 11.7% of the households. Other major effects included injuries (3.3%), collapse of buildings (3.3%) and loss of livestock/income generating activities (1.7%). The flood effects could have direct and indirect implications on the affected people (Jha et al. 2012), which can affect the physical, natural, social and economic wellbeing of affected members (Douglas et al. 2008). It seems to be true as the flood hazards have significantly influenced the lifestyle and economic conditions of the people. As 'heroic entrepreneurship' (as indicated by De Soto in his discussion on informality), Alajo serves as an economic space where individuals conduct their businesses such as vendors of vegetables, fruits and other food items, cobblers, barbers, second-hand clothes sellers, with many operating in small kiosks who are affected through flood events. These informal workers work without secure contracts and social protection (Debrah, 2007), as such occurrences of floods have profound economic implications on them. Again, the quality of life and well-being of households have been affected. This is because, Alajo also presents itself as informal settlement, depicting a true picture of its informality as both economic space and place of abode for people (see Werna, 2001; Porter et al., 2011). Due to the occurrence of flood events, many items of affected households such as mattresses, furniture etc become wet, which need to be washed, dried and re-used. This experience is unpleasant, and takes households' time away from other important activities, hence, reducing their quality of life and wellbeing. Quite apart from the socio-economic implications of the flood events, households have also been affected psychologically. This is particularly due to insecurity and the continuous effects experienced from the floods, which have translated to the innate local coping strategies to help adapt to floods.

> *"Anytime it rains and the gutters are getting full, we begin to panic. We have made our minds to leave here to a safe abode when we are financially ready"* **(Household head, woman).**

| Effects | Number of respondents | Percent value |
|---|---|---|
| Loss of household items | 41 | 68.3 |
| Outbreak of disease | 7 | 11.7 |
| Disruption and loss of services and businesses | 7 | 11.7 |
| Collapse of building | 2 | 3.3 |
| Personal injuries | 2 | 3.3 |
| Loss of livestock/income generating activity | 1 | 1.7 |
| Total | 60 | 100.0 |

*Table 1: Major effects of preceding floods on residents of Alajo*
*(Source: Authors' construct, 2018 based on field data)*

## 5.2 Local flood coping strategies

The rise in disasters directly connected to urban informality has negative effects on informal dwellers. This makes informal households to embrace '*indigenous*' measures to mitigate and respond to flood events. As indicated by Sakjege, Lupala and Sheuva (2012), informal communities are highly susceptible to flooding but lack support from formal state officials. Informality is closely associated with illegal use of lands for housing purposes, as it does not conform to planning regulations (UN-Habitat, 2003); and this makes it difficult for state officials to respond to its needs. In Alajo for instance, the detrimental effects of flood events characterized by the absence of state support have induced households to adopt '*indigenous*' means labelled as '*coping strategies*' in this paper to mitigate and reactively respond to flooding. These strategies are often short or long term efforts taken by individual households who were once victims of flood disaster(s) in order to survive the effects before another one, during the event or after the event has happened. Households in Alajo predominantly confirmed that despite their threats and insecurity, floods can be coped with when they occur.

One key local coping strategies through flood mitigation identified in the community was '*raised foundations of buildings*', which meant that the doors and windows of houses remain at a high level. This approach to cope with flood has been identified by Sakjege, Lupala and Sheuya (2011) in a related study in Tanzania. The use of this approach as iterated by households, ensures that flood waters cannot easily enter into rooms/inner parts of houses to destroy properties. They further indicated that the physiology of the terrain of Alajo allows for easy flow of water when drains are full, hence, making the use of such mitigation strategy a vital one. As victims of previous floods who have gained some considerable experiences, they asserted that the use of this approach has reduced the effects of flood, especially in relation to the destruction of their properties. In some households, sand bags have been used as a complementary method to block the overflow of water when there are rains. This approach has also been revealed by Adelekan (2010) and Amoako (2012) in their studies of urban informality and floods in Africa. During our field work, one household head (male) has this to say:

> "*We have realized that this place will always get flooded during the rainy season. So we decided to raise our buildings to higher heights after some flood encounters which destroyed many people's properties. This method has helped*

*us other than that, I am sure we will not have been here by now. When it rains and we have floods, the effects are not too massive. Also, some households have used sand bags to block the areas where the water normally flows from. I am very optimistic that next year when it floods, the effects will be very minimal on us".*

Apart from raising the foundations of their buildings and using sand bags, households have embraced proper disposal of wastes and clearing of gutters as a strategic response to flood hazards. Interactions with households indicated that poor drainage system in the community is a major cause of flood disaster. This coping strategy is usually adopted before and after some days of flood event. With this strategy, households indicated that they clear their respective drains, especially when the rainy season is approaching. This, according to them, ensures the free flow of water, hence, helping in the mitigation of floods. Households further confirmed that in some cases, residents who are very close to the two rivers transfer their expensive items to safe places, especially friends and relatives who live nearby. Others pack their items on shelves and other high levels when there are rains. Households also engage in the construction of temporal drains to channel water to appropriate locations, as an attempt to manage flood events. The Accra Metropolitan Assembly (AMA) has not played any role in that regard to support households in their risk mitigation measures. Hence, they are left to fence for themselves. These three aforementioned coping strategies are similar to what Douglas et al. (2008) identified in their study of flooding in Africa. Households added that they have local rescue teams formed, who assist households in the phase of floods as a form of reactive response to floods. The rescue teams entail young men who are locally recognized and contacted by households to help them for an agreed fee. The groups organize themselves in flood prone areas of Alajo during the rainy season to assist households to quickly transfer their properties to safer places.

> *"The rescue teams are made up of strong men who help us to move our properties to safe places when we perceive a possibility of flood occurring during the rainy season"* (reiterated by a Woman, household member).

In assessing the various flood coping strategies, a Likert scale of 'High', 'Medium' and 'Low' has been used to gage the reliability and cost of the strategies adopted by households. The choice of scale and the factors considered for the assessment were induced by available data solicited from the study participants. The results are shown in Table 2. The choice of adoption of a particular local coping strategy by households is highly influenced by the financial strength of the households. Hence, whilst a particular coping strategy could sound very dependable, the cost involved could scare households from embracing it. Majority of the households indicated that they wish they could pull down their buildings, and erect new ones with higher foundations, as this seems to work effectively for households who have embraced it. Unfortunately, the cost involved is high, hence, majority prefer to clear their gutters/drains and to pack their items on shelves or high levels as these methods are not too expensive.

| Local coping strategies | Scale for assessment | |
|---|---|---|
| | **Reliability** | **Cost** |
| By raising the foundation of building | 3 | 3 |
| By using sand bags to block water flows | 2 | 2 |
| By clearing choked gutters/drains | 2 | 1 |
| By early transfer of items to safe places | 2 | 2 |
| By packing items on shelves and high levels | 1 | 1 |
| By local rescue team | 1 | 2 |
| By constructing temporal drains | 2 | 2 |
| Parameter | Interpretation | |
| 1 | **Low** | |
| 2 | **Medium** | |
| 3 | **High** | |

*Table 2: Average results of reliability and cost of coping strategies:*

*Households' perspective*

**(Source: Authors' construct, 2018 based on field data)**

## 5.3 Determinants of local flood coping strategies

Households react to flooding through diverse local coping strategies in order to reduce risks. This is deem necessary as households living in flood zones will naturally try and reduce the impacts of flooding. However, the intensity of efforts to mitigate and response to flood could vary depending on differing factors pertaining to a particular flood prone community. Therefore, we contend that households' efforts to mitigate and response to flooding in Alajo are influenced by the following determinant factors of the community which somewhat influence the specific coping strategy/strategies embraced. The determinant factors are: i) the location of Alajo in-between two major rivers (locational disadvantage), (ii) the limited state support to the flood zones in the community; and iii) housing affordability and flexibility of property ownership which entice poor urbanites to stay in the community.

The locational disadvantage of Alajo has been unveiled by Douglas et al. (2008), who indicated that the location of the community automatically makes it a flood prone area in Accra. This indication goes hand in hand with Satterthwaite (2007) who revealed that informality can contribute to flooding when lower income groups settle on hazardous sites. We support this position as the visits to the community confirmed that the community is located in the midst of river Odaw and Onyasia (a hazardous site for flood occurrence). Households indicated that these rivers overflow their banks when there are heavy rains. Hence, households located very close to the rivers are influenced to deliberately raise the foundations of their buildings as the main coping strategy to deal with flooding. Households who are a bit distance from the river prefer to use sand bags and other materials to block the paths which have been previously identified to be the channels in which the rivers overflow to those households. Thus, the exact locations of households determine which intensive coping strategy they prefer to mitigate and respond to flooding.

31

*"Because I am too close to the Odaw river, I decided to allow my building to have high foundation so that when there are overflows, it does not destroy my properties"* (Household head, male).

*"My husband organized other households in the house to raise the blockage you find at the edge of the river. They loaded sand in sacks and bags as well as stones to block the path in which the overflow waters travel"* (Household member, female).

Discussions with households revealed that city authorities have failed to provide the necessary support and resources to support the community to manage and adapt to floods. They indicated that the state has not supported them in the provision and maintenance of drain infrastructure that contributes to risk reduction. Thus, building, maintaining and clearing drains have been the sole responsibilities of the informal urbanites with no state support. According to them, concerned state officials including those responsible for water and sanitation and disaster management have not assisted them in flood risks reduction. For instance, they indicated that the National Disaster Management Organization (NADMO), responsible for community-based preventive and management education, has failed to educate them on how best they can prevent and manage floods. The lack of formal state support has induced households to put in place informal coping strategies based on their own experiences to deal with the situation. National Disaster Management Organization (NADMO) officials confirmed that they are challenged with human, financial and logistical resources to effectively operate to support communities to prevent disasters. Discussions with officials from the Accra Metropolitan Assembly (AMA), a decentralized government body for local level development, confirmed that the

32

government only provides support in the event of floods (reactive response), but the support measures are inadequate due to limited funds. Hence, after flood disasters, relief items are normally provided by some private organizations such as Vodafone Ghana Limited and other philanthropic groups. The '*hand-outs*' from these non-state bodies are not enough, and also, it is not a precise way of dealing with floods. Households prefer precise and sustainable measures which will safeguard their safety. Government, as revealed by the National Disaster Management Organization (NADMO), is not well positioned to face the full financial burden of relocating community members from Alajo. Again, apart from eviction being a cruel means to deal with the flood, it will not be appropriate as some informal dwellers will later on find their way back to the community (reiterated by a city planner at Town and Country Planning Department (TCPD)). With government not too sure about how to deal with the flood events, households in Alajo are determined to use their little resources to cope differently with flooding in tandem with their divergent experiences with flood events in the community.

In Osei and Ampratwum (2014) study, they identified housing affordability and flexibility of property ownership in informal Accra as a major reason why people prefer to stay in such locations. Our field work confirmed that rent and property cost in Alajo require very low financial resources. Average rent for instance, is 35.00 Ghana cedis per month as compare to formal Accra where average rent is at minimum 120 Ghana cedis. Apart from the low cost of living in Alajo, the community is strategically located closer to the CBD of Accra, which presents a pool of market for business activities (a hub with concentration of informal economic activities). Majority of households in Alajo engage in informal activities at the CBD. These household members cannot afford the high cost of rent and/shelter in the CBD, and as such prefer to stay in Alajo which is about six (6) kilometers from the CBD. Some households also indicated that they have made investment in acquiring land and other properties in Alajo, and as such, find it difficult to abandon them. Others also confirmed that they have their businesses and sources of livelihood at the community and are not financially ready to move to a safer community despite threat. Though they claimed they will move once they are financially prepared, there are no indications to confirm to the move at any moment from now. Some households live in improvised buildings such as kiosks, containers and uncompleted buildings which lack basic water and sanitation facilities such as potable drinking water and toilet facilities. Satterthwaite et al (2011) identified that the urban poor live in slums and other informal areas due to housing deficit and the high cost of rent in the cities; a revelation which certainly hold true for Alajo, a slum community. With these compelling factors influencing their stay, they have been induced to embrace local coping strategies to face flood, which is the major threat to their stay in the community.

## 5.4 Examining the coping strategies and their contributions to the adaptive resilience of households using the DROP model

We start by looking at the informal nature of Alajo in terms of its social systems, natural systems and the physical/built environment, and how they have influenced the degree of inherent vulnerability and inherent resilience of households to flood. Taking into account the social systems, Alajo is a cosmopolitan community, compose of mainly Ga (indigenes), Mole-Dagbons (migrants from the north), Ewes and Akans (migrants from the south) as ethnic groups (Atuguba and Amuzu, 2006). The community has high poverty levels, and its population persistently keeps growing due to its strategic location. The coping measures to flood in the community is mainly household-based instead of entire community's initiatives. Generally, the physical development of the community is haphazardly organized. Individual households do not coordinate in their efforts to manage flood events, instead, they examine specific situations of their physical

34

environment, and find appropriate coping strategy/strategies that will help them to specifically manage and adapt to the events. For instance, we realized that households living closer to the two rivers have raised foundations of their buildings, whilst those closer to major drains clear the drains to allow for easy flow of water especially, during rainy seasons. The lack of local coordination in coping with flood contributes to very little success in local flood adaptation in informal communities (Amoako, 2012), and this seems to hold true for Alajo. Though the natural systems (in terms of nature of land and location) directly expose the community to flood (as seen as a feature of urban informality as presented by Elgin and Oztunali, 2012; Satterthwaite et al., 2011), households have failed to team up to build a strong social system for a supportive physical/built environment which can help them to adapt successfully to flood and subsequently build their resilience to flood over time.

Inferring from the DROP-model, the community can be said to have high degree of inherent vulnerability and low degree of inherent resilient, as antecedent conditions of social systems, natural systems and built environment are unfavorable. The immediate effects of preceding floods (loss of household items, disruption and loss of services and businesses, personal injuries etc) are influenced by the floods event characteristics (duration, frequency, intensity, magnitude and rate of its onset). The floods' effects can be reduced through coping strategies (Cutter et al., 2008). The coping strategies of households (raised foundations of buildings, the use of sand bags to block water flows, cleared choked drains, early transfer of items to safe places, items packed on shelves and high levels, local rescue team engagements and construction of temporal drains) have yielded some positive results, as households claimed the effects of flood hazards have gradually reduced. This is a positive indication (+) and has direct positive outcome on the absorptive capacity of households as well as their resilience. Over the years, households have implemented lessons learnt from preceding floods and this has underpinned their differing coping strategies. However, social learning in flood events has been absent, and this makes the absorptive capacity of Alajo very weak. Households revealed that the coping strategies they embrace are different from one another depending on the specific locations (either closer to the two major rivers, a bit distant from the rivers or at areas closer to major drains) in the community. Households tend to implement coping measures without teaming up with other households, especially, those in different houses. Hence, there are no coordination of efforts, knowledge sharing and resource mobilization to holistically tackle peculiar issues which could be handled by households to deal with floods. This clearly shows the absence of social learning in interventions put in place. Lessons are learnt at individual household's level through their own experiences with preceding floods, but these experiences are not shared among affected households to inform sound coping measures needed for strong adaptive capacity for high adaptive resilience to both minor and major flooding.

> *"Here in Alajo, we all have our own way of dealing with flooding. I have never seen any special gathering instituted by our leaders for us to share ideas on how best to deal with the situation. Members of every house cope with the situation through their own understanding and means which they believe will be enough to mitigate or respond to the event"* (Household Head, man)

Adger et al. (2005) have indicated that social learning in community-based disaster prevention and/or mitigation is very central, and should be embraced by all responsible bodies and individuals so as to tackle disaster events and their detrimental outcomes. They presented that social learning depends on the ability of households to embrace and build strong social integration which allows for group-based actions to deal with a disaster such as flood. This subsequently feeds into the absorptive capacity of households making them more resilient to flood. Due to the absence of social learning, we have doubt in the absorptive capacity of households in Alajo, but, we do admit that the coping strategies of households (informed by lessons learnt but lack social learning) can help them to adapt to minor flood events. In such situations, the absorptive capacity of households will not be exceeded, and their degree of recovery will be very high. In the phase of major flood event, the absorptive capacity of the community will be exceeded and the degree of recovery will as such be very low. Thus, Alajo has high adaptive resilience to minor floods with high degree of recovery but low adaptive resilience to major flood with low degree of recovery (See Figure 6).

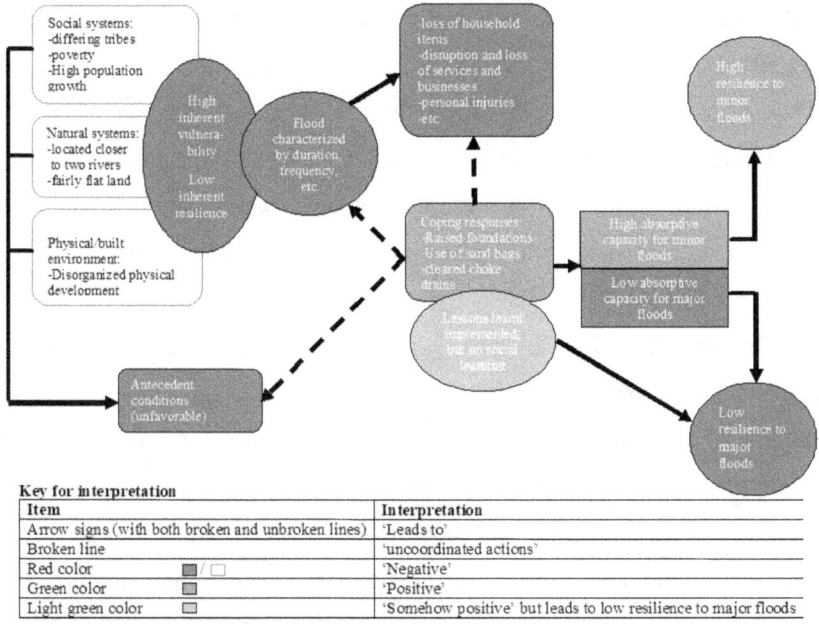

Key for interpretation

| Item | | Interpretation |
|---|---|---|
| Arrow signs (with both broken and unbroken lines) | | 'Leads to' |
| Broken line | | 'uncoordinated actions' |
| Red color | ▨ / ☐ | 'Negative' |
| Green color | ▨ | 'Positive' |
| Light green color | ☐ | 'Somehow positive' but leads to low resilience to major floods |

*Figure 6: Conceptualized DROP-model for Alajo*

**(Source: Authors' construct, 2018 based on literature and field data)**

# 6. CONCLUDING REMARKS: TOWARDS EFFECTIVE FLOOD MANAGEMENT IN THE CONTEXT OF ALAJO WITH IMPLICATIONS FOR EFFECTIVE URBAN POLICY DESIGN

Households have taken deliberate steps to cope with flooding in Alajo. The coping strategies have yielded some positive outcomes as confirmed by households. These measures, however, need to be concretized to ensure effective local adaptation leading to a state of resilience. This is particularly relevant as the strategies implemented by households over the past years have contributed little to building resilience to flooding. First, we recommend households to take a collaborative approach to handling flood issues, through knowledge sharing, and coordinated efforts to help them advance the reliability of their local coping strategies. Social learning is thus, one key aspect households should embrace to build on their adaptive capacity to flooding. Second, we call on the state not to evict the informal urbanites, instead, locate them in their informality and support their '*indigenous*' strategies for profound outcomes. Obviously, informality has become part of our cities, and as such, we do not expect city authorities to out-rightly tag it as a problem (as also argued by: Roy and Al-Sayyad, 2004). Thus, city authorities should offer their support which could also include flood mitigation and adaptation training programs offered to some community members and groups. Such individuals can later become local pioneers and flood control '*ambassadors*' who will serve as a channel of exchanges between formal state institutions and the informal urbanites. Also, throughout the research, we realized that households' coping strategies did not consider rainwater harvesting. Therefore, we contend that the adoption of rainwater harvesting by households in Alajo could be very helpful in managing flooding. This is due to the low lying nature of the community which allows for run offs from rains leading to erosion, and ultimately flooding. Apart from the recommendations for local flood

38

management, we wish to make theoretical contribution to the DROP model. In as much as the DROP model has been internalized for community based disaster management, we call for the model's inculcation of '*external supports*' as a measure in concretizing coping strategies for building local resilience to flood. Apart from '*social learning*' which tends to increase the adaptive resilience of communities, '*external supports*' which are commensurate to managing disasters also contribute to the transformation of disaster vulnerability communities

**ACKNOWLEDGEMENT:** The authors thank the Almighty God, the informal urbanites who participated in this research, officials of the state institutions who were engaged and the anonymous reviewers who provided their inputs to shape the article.

## LITERATURE:

Aboagye, D. (2012). The political ecology of environmental hazards in Accra, Ghana. *Journal of Environment and Earth Science*, 2(10), 157-172.

Accra Metropolitan Assembly (AMA) (2014). The Medium Term Development Plan, 2010-2015, Accra, Ghana

Adegun, O. B. (2011). Shelter and the future African city. *The Built & Human Environment Review*. 4 (2), 33-40.

Adelekan, O. I. (2010). 'Vulnerability of poor urban coastal communities to flooding in Lagos, Nigeria', *Environment and Urbanization*, Vol. 22 no. 1, pp. 433-450

Adger W.N, Hughes T., Folke C., Carpenter S. and Rockström J. (2005). Social-ecological resilience to coastal disasters. *Science* 309:1036 –1039

Adjei-Mensah, C., Antwi, K.B. and Acheampong, P.K. (2013). Behavioural Dimension of the Growth of Informal Settlements in Kumasi city, Ghana, *Research on humanities and social sciences*, Vol. 3 (12), pp. 1-10

Amoako, C. (2012). Emerging issues in urban flooding in African cities-The case of Accra, Ghana, 35th AFSAAP Annual Conference Proceeding 2012-www.afsaap.org.au

Amoako, C. (2014). The Politics of Flood Vulnerability in Informal Settlements around the Korle Lagoon in Accra, Ghana. *OPPORTUNITIES*, 50.

Attipoe, S. K. (2015). *An assessment of flood mitigation measures in Accra, Ghana* (Doctoral dissertation).

Atuguba, R. A. and Amuzu, T. E. (2006). Report on Climate Change and Flooding in Alajo, Accra. The Legal Resources Center-Ghana.

Boubacar, S., Pelling, M., Barcena, A. and Montandon, R. (2017). The erosive effects of small disasters on household absorptive capacity in Niamey: a nested HEA approach. *Environment and Urbanization*, 29(1), pp.33-50.

Birkmann, J. (2011). Regulation and coupling of society and nature in the context of natural hazards. In *Coping with Global Environmental Change, Disasters and Security* (pp. 1103-1127). Springer, Berlin, Heidelberg.

Braun, B. and Aßheuer, T. (2011). Floods in megacity environments: vulnerability and coping strategies of slum dwellers in Dhaka/Bangladesh. *Natural hazards*, 58(2), pp.771-787.

Brooks, N., Adger, W.N. and Kelly, P.M. (2005). The determinants of vulnerability and adaptive capacity at the national level and the implications for adaptation. *Global environmental change*, *15*(2), pp.151-163.

Carpenter S, Walker B, Anderies J.M, and Abel N. (2001). From metaphor to measurement: resilience of what to what? *Ecosystems.* 4:765–781. Doi: 10.1007/s10021-001-0045-9

Creswell, J. (2012). Educational Research: Planning, Conducting, and Evaluating Quantitative and Qualitative Research (4th ed.). Boston, MA: Pearson Education, Inc.

Creswell, J.W. (2014). Research design: Qualitative, quantitative, and mixed methods approaches. Sage

Cutter, S.L., Barnes, L., Berry, M., Burton, C., Evans, E., Tate, E. and Webb, J. (2008). A placed-based model for understanding community resilience to natural disasters, *Global environmental change*, 18:598-606

De Risi, R. (2013). A probabilistic bi-scale framework for urban flood risk assessment. PhD dissertation, Department of Structures for Engineering and Architecture, University of Naples Federico II, Naples

De Soto, H. (2000). *The mystery of capital: why capitalism triumphs in the West and fails everywhere else*. New York: Basic Books

De Soto, H. (1989). *The other path: The invisible revolution in the Third World.* London: I.B. Taurus.

41

Debrah, Y. (2007). Promoting the Informal Sector As A Source Of Gainful Employment In Developing Countries: Insights From Ghana. *The International Journal of Human Resource Management* 18(6): 1063-1084

Douglas I., Alam K., Maghenda M., Mcdonnell Y., Mclean L. and Campbell, J. (2008). *Unjust Waters: Climate Change, Flooding and the Urban Poor In Africa.* International Institute for Environment and Development (IIED).

Duminy, J. (2011). Literature survey: Informality and planning. *Being literature survey complied for the Urban Policies Programme of the policy-research network Women in informal Employment: Globalizing and Organizing (WIEGO)*

Dörnyei, Z. (2007). Research methods in applied linguistics. New York: Oxford University Press

Elgin, C. and Oztunali, O. (2012). Shadow economies around the world: model based estimates. *Bogazici University Department of Economics Working Papers*, 5, pp.1-48.

Fekade, W. (2000). Deficits of formal urban land management and informal responses under rapid growth, an international perspective. *Habitat International* 24/2 pp 127-150

Food and Agricultural Organization (FAO) (2012). Building resilience adaptation to climate change in the agricultural sector. Proceedings of a Joint FAO/OECD Workshop 23-24 April, 2012, Rome

Ghana Statistical Service (GSS) (2012). 2010 Population and Housing Census. Summary Report of Final Results, Accra, Ghana.

Gow, G. A., 2005. Policymaking for Critical Infrastructure. Ashgate, Alder-shot

Guha-Khasnobis, B., Kanbur, R., and Ostrom, E. (Eds.) (2006). *Linking the formal and informal economy: concepts and policies.* Oxford University Press.

Hall, P. and Pfeiffer, U. (2000). *Urban future 21: A global agenda for 21$^{st}$ century cities, London*: E & FN Spon

Heintz, J. (2012). Informality, inclusiveness, and economic growth: an overview of key issues. *International Development Research Centre (IDRC)*.

Hollohan, R. and Barry, C. (2014). Mixed Methods: Focus on Convergent Parallel Design (a presentation available at: https://prezi.com/j9sa5hyawlop/mixed-methods-convergent-parallel/

Jha, A. K., Bloch, R. and Lamond, J. (2012). *Cities and flooding: a guide to integrated urban flood risk management for the 21st century*. World Bank Publications.

Kasperson, J.X., Kasperson, R.E., Turner, B.L., Hsieh, W. and Schiller, A. (2012). Vulnerability to global environmental change. In *The Social Contours of Risk: Volume II: Risk Analysis, Corporations and the Globalization of Risk*. Taylor and Francis.

Liverman, D. M. (1990). Vulnerability to global environmental change. In: Kasperson, R. E., Dow, K., Golding, D., Kasperson, J. X. (Eds.), Under-standing Global Environmental Change: The Contributions of Risk Analysis and Management. Clark University, Worcester, MA, Ch. 26, pp. 27-44.

Morse, J.M. (2003). Principles of mixed methods and multimethod research design. *Handbook of mixed methods in social and behavioral research, 1*, pp.189-208.

O'Brien, K., Eriksen, S., Schjolen, A. and Nygaard, L. (2004a). What's in a word? Conflicting interpretations of vulnerability in climate change research. CI-CERO Working Paper 2004:04, CICERO, Oslo University, Oslo, Norway

Osei-Boateng, C. and Ampratwum, W. (2014). The informal sector in Ghana. FES-Ghana. *Accessed on 10th September*.

Owusu, G. and Afutu-Kotey, R. L. (2010). Poor urban communities and municipal interface in Ghana: A case study of Accra and Sekondi-Takoradi metropolis. *African Studies Quarterly, 12* (1), 1.

Pelling, M. (2010). *Adaptation to climate change: from resilience to transformation*. Routledge.

Porter, L., Lombard, M., Huxley, M., Ingin, A.K., Islam, T., Briggs, J., Rukmana, D., Devlin, R. and Watson, V. (2011). Informality, the commons and the paradoxes for planning: Concepts and debates for informality and planning self-made cities: Ordinary informality? The reordering of a Romany neighborhood the land formalization process and the peri-urban zone of Dar es Salaam, Tanzania street vendors and planning in Indonesian cities informal urbanism in the USA: New challenges for theory and practice engaging with citizenship and urban struggle through an informality lens. *Planning Theory & Practice, 12*(1), pp.115-153.

Roy A. and Al-Sayyad, N. (Eds) (2004). *Urban informality: Transnational perspectives from the Middle East, South Asia and Latin America*. Lanham, MD: Lexington Books

Sakijege, T., Lupala, J., and Sheuya, S. (2012). Flooding, flood risks and coping strategies in urban informal residential areas: The case of Keko Machungwa, Dar es Salaam, Tanzania. *Jàmbá: Journal of Disaster Risk Studies, 4*(1), 1-10.

Satterthwaite D., Mitlin D. and Patel S. (2011). *Engaging with the Urban Poor and their Organizations for Poverty Reduction and Urban Governance*. An issues paper for the United Nations Development Programme.

Dodman, D., Bicknell, J. and Satterthwaite, D. (2012). Adapting to Climate Change in Urban Areas: The Possibilities and Constraints in Low-and Middle-Income Nations: David Satterthwaite, Saleemul Huq, Hannah Reid, Mark Pelling and Patricia Romero Lankao. In *Adapting Cities to Climate Change* (pp. 25-69). Routledge.Songsore, J. (2008). Environmental and structural inequalities in Greater Accra. *Journal of the International Institute, 16*(1).

Songsore, J. (2009). The Urban Transition in Ghana: Urbanization National Development and Poverty Reduction; thesis submitted to the Department of Geography and Resource Development, University of Ghana, Legon-Accra; Ghana

Tschakert, P., Sagoe, R., Ofori-Darko, G. and Codjoe, S.N. (2010). Floods in the Sahel: an analysis of anomalies, memory, and anticipatory learning. *Climatic Change*, *103*(3-4), pp.471-502.

Turner II, B. L., Kasperson, R. E., Matson, P. A., McCarthy, J. J., Corell, R. W., Christensen, L., Eckley, N., Kasperson, J. X., Luers, A., Martello, M. L., Polsky, C., Pulsipher, A., Schiller, A. (2003). A framework for vulnerability analysis in sustainability science. Proceedings of the National Academy of Sciences of the United States of America 100, 8074{8079.

Twumasi-Ankrah K. (1995). Rural-Urban Migration and Socioeconomic Development in Ghana: Some Discussions. *J. Soc. Devel. Afri*, 10(2): 13-22.

UN-Habitat (2003). Global Report on Human Settlements 2003: The Challenge of Slums. *London: Earthscan*.

UN-habitat (2010). *State of the world's cities 2010/2011: bridging the urban divide*. EarthScan.

UN-Habitat (2011). Participatory slum upgrading and prevention millennium city of Accra, Ghana. *Accra*

United Nations (2004). World population policies 2003. New York. USA. United Nations

United Nations Centre for Human Settlement (UNCHS) (2007). Urbanization: A Turning Point in History. Global Report on Urbanization [Online] Available: www.unhabitat.org.

United Nations Population Division (2002). World Population Prospects: The 2002 Revision, Highlights (online database). ESA/P/WP.180,United

Nations 2004. World population policies 2003. New York. USA. United Nations.

Werna, E. (2001). Combating urban inequalities. Challenges for managing cities in the developing world.

Wittink, M.N., Barg, F.K. and Gallo, J.J. (2006). Unwritten rules of talking to doctors about depression. Integrating qualitative and quantitative methods. *Annlas of Family medicine*, 4, 302-309, doi:10"1370/afm.558

Yin, R. K. (2003). Case study research: design and methods, Applied social research methods series. *Thousand Oaks, CA: Sage Publications, Inc. Afacan, Y., & Erbug, C.(2009). An interdisciplinary heuristic evaluation method for universal building design. Journal of Applied Ergonomics, 40,* 731-744.

Yohe, G. and Tol, R.S. (2002). Indicators for social and economic coping capacity—moving toward a working definition of adaptive capacity. *Global environmental change, 12*(1), pp.25-40.

Publisher: Eliva Press SRL

Email: info@elivapress.com

**Eliva Press** is an independent publishing house established for the publication and dissemination of academic works all over the world. Company provides high quality and professional service for all of our authors.

Our Services:
Free of charge, open-minded, eco-friendly, innovational.

-All services are free of charge for you as our author (manuscript review, step-by-step book preparation, publication, distribution, and marketing).
-No financial risk. The author is not obliged to pay any hidden fees for publication.
-Editors. Dedicated editors will assist step by step through the projects.
-Money paid to the author for every book sold.  Up to 50% royalties guaranteed.
-ISBN (International Standard Book Number). We assign a unique ISBN to every Eliva Press book.
-Digital archive storage. Books will be available online for a long time. We don't need to have a stock of our titles. No unsold copies. Eliva Press uses environment friendly print on demand technology that limits the needs of publishing business. We care about environment and share these principles with our customers.
-Cover design. Cover art is designed by a professional designer.
-Worldwide distribution. We continue expanding our distribution channels to make sure that all readers have access to our books.

**www.elivapress.com**